MASTERING THE MOVE: ESSENTIAL SCRIPTS FOR SUCCESSFUL FREIGHT BROKERS

TABLE OF CONTENTS

INTRODUCTION

FREIGHT BROKERAGE SERVICES

In the dynamic world of commerce and industry, where the gears of global trade turn incessantly, the seamless movement of goods is nothing short of the lifeblood that courses through the veins of our modern economy. It is this relentless pulse that sustains the intricate web of global commerce, ensuring that businesses can thrive and consumers can access the products they desire. Within this intricate ecosystem, freight brokerage services emerge as the unsung heroes who orchestrate this complex symphony of shipping, harmonizing the cacophony of logistical challenges, and transforming them into triumphant logistical feats.

Much like the conductors of a well-orchestrated symphony, freight brokers possess an uncanny ability to wield their expertise and networks to facilitate the efficient flow of goods. They are the invisible hand that connects businesses, carriers, and consumers, weaving together the threads of trade into a seamless tapestry of commerce. In this comprehensive exploration, we'll journey deep into the fascinating realm of freight brokerage services, shedding light on the multifaceted world that thrives behind the scenes, out of the spotlight, and yet plays a pivotal role in driving the global economy forward.

As we embark on this journey, fasten your seatbelt, for the world of freight brokerage is a fast-paced and ever-evolving landscape where precision, adaptability, and a keen eye for opportunity are the guiding stars. Within this realm, freight brokers navigate through a labyrinth of choices, helping businesses make critical decisions about how to transport their goods. It's a world where time is of the essence, where efficiency translates directly into savings, and where a misstep can disrupt supply chains and ripple across industries.

The importance of freight brokerage services becomes even more evident when we consider the vast array of goods and products that traverse the globe each day. Whether it's fresh produce making its way from distant farms to supermarket shelves, high-tech gadgets moving from manufacturers to eager consumers, or raw materials needed to keep industries humming, freight brokers are the silent architects of these intricate logistical maneuvers.

Yet, the role of a freight broker extends far beyond simply arranging transportation. These experts in the art of logistics are responsible for negotiating rates, tracking shipments, managing documentation, and mitigating risks. They are the glue that holds together the global supply chain, ensuring that the wheels of commerce keep turning smoothly.

At the heart of their responsibilities is the task of connecting shippers and carriers. With an extensive network of carriers specializing in various transportation modes—trucking, rail, air, and sea—freight brokers possess an encyclopedic knowledge of the available options. This enables them to make informed decisions about the best carriers to transport specific types of cargo, covering various distances and meeting delivery deadlines. In essence, they are the matchmakers of the shipping world, ensuring that every load finds its ideal carrier partner.

Negotiating rates is another critical aspect of their role. They must be astute in reading market conditions, analyzing shipment volumes, and factoring in various variables to secure the best rates for their clients. This requires a keen understanding of the ebb and flow of the shipping industry, as well as the ability to strike a balance between cost-efficiency for the shipper and fair compensation for the carrier.

In the intricate world of shipping, where numerous variables can throw even the best-laid plans off course, freight brokers are masters of problem-solving. They leverage technology to track shipments in real-time, allowing them to anticipate and address issues before they become disruptions. This not only ensures that products reach their destinations on time but also provides shippers with valuable visibility into the movement of their goods, giving them peace of mind and greater control over their supply chains.

The domain of freight brokerage services isn't merely a web of relationships, negotiations, and logistics. It's also a landscape where the ever-advancing tide of technology is reshaping the way these professionals operate. Digital platforms and automation tools have revolutionized the industry, streamlining processes, enhancing transparency, and improving the overall efficiency of freight brokerage. This evolution has enabled brokers to provide even more value to their clients, as they can access real-time data and insights that were previously unavailable.

But it's not just about technology; the human touch remains an essential element in this industry. Freight brokers bring not only their wealth of experience and industry knowledge but also their interpersonal skills to the table. The relationships they cultivate, which can span years or even decades, form the bedrock of trust and reliability in an industry that relies on partnerships to function seamlessly.

In conclusion, freight brokerage services are the often-overlooked linchpin of the global economy, where businesses, carriers, and consumers intersect. They ensure that the flow of goods remains uninterrupted, cost-efficient, and reliable. When freight brokers orchestrate the logistics, the world's markets flourish, and the heartbeat of global trade remains strong, propelling commerce forward into a future where goods continue to traverse the globe, creating new opportunities and connections. So, as we venture further into this world, be prepared to be amazed by the intricate choreography that keeps our modern economy in motion

.

CHAPTER ONE

"COLD SCRIPTS FOR BROKERS"

"Script" is a strategic framework, typically a written document or verbal guide, meticulously crafted to facilitate effective communication between brokers and potential clients. This script is an essential tool in the broker's toolkit, providing a structured approach to initiate, navigate, and conclude conversations that lead to successful partnerships and transactions.

It encompasses a well-planned sequence of dialogue, including introductions, key talking points, objection handling, and calls to action. By following this script, brokers can confidently convey the unique value they bring to the table, address clients' concerns, and ultimately, secure their trust and business. A well-designed script not only streamlines the broker-client interaction but also highlights the broker's expertise and the benefits of their services, ensuring that clients are well-informed and confident in their decision-making.

Cold scripts for brokers are carefully crafted communication templates designed to initiate contact with potential clients or prospects in a concise and effective manner. These scripts serve as a strategic tool for brokers to navigate the initial stages of client acquisition by presenting key information about their services and establishing a connection. Typically tailored to highlight the broker's expertise, unique selling propositions, and the value they bring to clients, these cold scripts aim to capture the recipient's attention and prompt further engagement.

Effective cold scripts for brokers often incorporate elements of personalization, demonstrating an understanding of the client's needs and concerns. By striking a balance between professionalism and relatability, brokers can create a positive first impression and lay the foundation for a meaningful business relationship. In a competitive market, the ability to articulate a compelling narrative through cold scripts can be a pivotal factor in building a client base and fostering long-term success in the brokerage industry.

Hello, [Prospect's Name],

I hope this message finds you well. I'm [Your Name] from [Your Company], and I wanted to talk to you about how our freight brokerage services can revolutionize your shipping operations. We understand that managing freight can be a complex and time-consuming task, but with our expertise and resources, we can simplify the process and save you time and money.

❖ **Establishing Rapport:**

➢ "Before we dive into the details, I'd love to learn more about your current shipping challenges and goals. Could you share some insights with me?"

❖ **Presenting the Problem:**

➢ "Many businesses face logistical challenges that lead to increased costs, delays, and operational inefficiencies. Coordinating shipments, finding reliable carriers, and ensuring on-time deliveries can be a daunting task."

❖ **Introduce Your Solution:**

➢ "That's where we come in. Our freight brokerage services are designed to make your life easier and your logistics more efficient. We act as your strategic partner, connecting you with the right carriers and optimizing your supply chain."

- ❖ **Highlight Key Benefits:**
- ➢ **Cost Savings:** "Our extensive network of carriers allows us to negotiate competitive rates, saving you money on every shipment."
- ➢ **Time Efficiency:** "We handle the logistics, so you can focus on what you do best – growing your business."
- ➢ **Reliability:** "We work with a vetted network of carriers, ensuring that your shipments arrive on time and in perfect condition."
- ➢ **Scalability:** "Whether you have a small or large shipment volume, we can adapt to your needs and help your business grow."
- ➢ **Provide Examples or Case Studies:** "Let me share a success story from one of our clients [Client Name]. They saw a [percentage] reduction in shipping costs and a [percentage] increase in on-time deliveries within just [time frame]."

- ❖ **Address Potential Objections:**
- ➢ "I understand you might have concerns. However, we have a dedicated team of experts who handle any issues that may arise, so you can trust us to deliver on our promises."

❖ **Call to Action:**

➤ "Would you be interested in exploring how our freight brokerage services can benefit your business? I'd love to schedule a brief call or meeting to discuss your specific needs."

❖ **Overcome Objections:**

➤ If the prospect hesitates, ask them what their concerns are and address them directly.

❖ **Summarize:**

➤ "In summary, partnering with [Your Company] can save you time, money, and headaches when it comes to managing your freight. We're here to make shipping easier for you."

❖ **Close the Deal:**

➤ "When can we schedule a follow-up conversation to delve deeper into your logistics needs and explore how our services can be tailored to your business?"

❖ **Thank the Prospect:**

➤ "Thank you for your time, [Prospect's Name]. I appreciate the opportunity to discuss how [Your

Company] can make a positive impact on your shipping operations."

❖ **Follow Up:**
➢ Send a follow-up email or message to express your continued interest and willingness to assist.

Remember to personalize your script to the specific needs and concerns of your prospect. Tailor your approach based on their industry, size, and any unique challenges they may face in their shipping operations.

Introduction:

[Start with a warm and professional greeting]

Salesperson: "Hello! My name is [Your Name], and I'm a freight broker with [Your Company Name]. I hope you're having a great day. May I have a moment of your time to discuss how our services can benefit your business?"

❖ **Establishing Rapport:**

[Build a connection with the prospect]

➢ **Salesperson**: "Before we dive into the details, can you tell me a bit about your current shipping and logistics needs? What challenges are you facing in managing your freight shipments?"

[Listen actively to their response and show genuine interest in their concerns.]

❖ **Highlighting Expertise:**

[Establish your company's credibility]

➢ **Salesperson:** "At [Your Company Name], we've been helping businesses like yours streamline their logistics operations for over [X] years. Our team of experienced freight brokers understands the industry inside out, and we have a track record of delivering cost-effective and efficient solutions."

❖ **Identifying Needs:**

[Ask questions to uncover their specific needs]

➢ **Salesperson:** "To better assist you, could you share some details about the types of products you ship, your typical shipment volumes, and any key destinations or challenges you face?"

❖ **Presenting Benefits:**

[Highlight how your services can solve their problems]

- ➢ **Salesperson:** "Our services are designed to make your logistics more efficient and cost-effective. Here are some of the key benefits we can offer:
- ➢ **Cost Savings:** We have strong relationships with carriers, allowing us to negotiate competitive rates on your behalf.
- ➢ **Reliability:** Our network of trusted carriers ensures timely deliveries and reduces the risk of shipment delays.
- ➢ **Time Savings:** We handle all the logistics paperwork, freeing up your time to focus on other aspects of your business.
- ➢ **Visibility:** Our tracking and reporting systems provide real-time visibility into your shipments.
- ➢ **Customized Solutions:** We tailor our services to meet your specific needs, whether it's full truckload, less than truckload, intermodal, or specialty freight."

❖ **Handling Objections:**

[Anticipate common objections and address them]

- ➢ **Salesperson:** "I understand you might have concerns about outsourcing your freight management. Let me address some common questions:

- ➤ **Control:** You maintain control over your shipments, and we work closely with you to meet your requirements.
- ➤ **Cost:** Our goal is to reduce your overall shipping costs through our industry knowledge and negotiation skills.
- ➤ **Reliability:** We have a stringent carrier selection process to ensure your shipments are in safe hands.
- ➤ **Communication:** We keep you informed throughout the shipping process, providing updates as needed."
- ➤ **Closing:**

[Ask for their commitment]

- ➤ **Salesperson:** "Based on our discussion, it seems that our services align well with your logistics needs. Are you ready to take the next step and explore how we can work together to optimize your freight management?"
- ➤ **Handling the Response:**

[Respond to their decision, whether it's a positive commitment, a need for more information, or a rejection]

Salesperson: "Great! I'll have one of our experts reach out to you shortly to discuss the next steps.

If you need more information or have further questions, please feel free to ask. We're here to help you streamline your logistics and improve your bottom line."

Remember to customize this script to fit your specific company's services and the needs of your potential clients. Adapt it as necessary and practice your delivery to ensure a confident and effective sales pitch.

❖ Introduction:

➤ **Agent:** "Hello, [Prospect's Name]. My name is [Your Name], and I represent [Your Company], a trusted freight brokerage service. How are you today?"

➤ **Prospect:** [Response]

➤ **Agent:** "Great to hear! I wanted to reach out to you because I believe we can help optimize your shipping and logistics operations while saving you time and money. Would you be open to a brief conversation to explore how we can assist you?"

❖ Establishing Rapport:

➤ **Agent:** "[Prospect's Name], we work with businesses just like yours, and we've been able to help them achieve significant cost savings and efficiency improvements. Can you share a bit about your current shipping and logistics challenges or goals?"

- ➢ **Prospect:** [Response]

- ❖ **Presenting Value:**

- ➢ **Agent:** "I understand your challenges, and that's where we excel. Here are some key benefits of partnering with [Your Company]:

- ➢ **Cost Savings:** We have access to a vast network of carriers and can negotiate competitive rates on your behalf, potentially reducing your shipping costs.

- ➢ **Time Savings:** We handle all the logistics coordination, paperwork, and tracking, so you can focus on your core business.

- ➢ **Reliability:** Our experienced team ensures that your shipments arrive on time and in good condition, reducing disruptions in your supply chain.

- ➢ **Scalability:** Whether you're a small business or a large corporation, we can tailor our services to meet your specific needs.

- ➢ **Technology:** We utilize advanced tracking and reporting systems to provide real-time visibility into your shipments.

❖ **Building Interest:**

- ➢ **Agent:** "[Prospect's Name], can you see how these benefits might address some of the challenges you mentioned earlier?"

- ➢ **Prospect:** [Response]

❖ **Handling Objections:**

- ➢ **Agent:** "I understand your concerns. Many of our clients initially had similar reservations, but after partnering with us, they've seen substantial improvements. We also offer a trial period to demonstrate our capabilities, and there are no long-term commitments required."

❖ **Closing the Sale:**

➢ **Agent:** "Given what we've discussed, would you be interested in scheduling a more in-depth consultation or a demonstration of our services? It would be an excellent opportunity to explore how we can tailor our solutions to fit your specific needs."

➢ **Prospect**: [Response]

❖ **Follow-up:**

➢ **Agent:** "Thank you, [Prospect's Name], for your time today. I'll send you an email with some additional information and available times for a follow-up call. I look forward to the possibility of working together to streamline your logistics and reduce your shipping costs."

Remember that successful sales calls often require flexibility and adaptability, so be prepared to adjust your script based on the prospect's responses and specific needs. Good luck with your freight brokerage sales!

❖ Introduction:

[Begin with a warm and professional greeting]

"Good [morning/afternoon], [Prospect's Name]. My name is [Your Name], and I represent [Your Company Name], a leading freight brokerage firm. I hope you're doing well today."

➤ Establishing Credibility:

"We've been in the logistics industry for [X] years, and our team has a proven track record of helping companies like yours streamline their supply chain, reduce costs, and improve efficiency. I wanted to reach out to S today because I believe we can add significant value to your business."

➤ Understanding Their Needs:

"To better assist you, could you please share some information about your current freight and logistics challenges? Are there specific pain points or goals you'd like to address?"

❖ **Presenting Solutions:**

"Based on what you've shared, I believe we have several solutions that could benefit your company. Here are some of the ways we can help:

➢ **Cost Savings:** Our extensive network of carriers and industry expertise allows us to negotiate competitive rates, potentially reducing your shipping costs.

➢ **Efficiency:** We can optimize your shipping routes and schedules to minimize transit times and maximize on-time deliveries.

➢ **Reliability:** With our tracking and monitoring systems, you'll have real-time visibility into your shipments, ensuring peace of mind and timely deliveries.

➢ **Customized Services:** We tailor our services to meet your unique requirements, whether it's LTL, FTL, intermodal, or specialized freight.

❖ Value Proposition:

"At [Your Company Name], we pride ourselves on delivering not just transportation solutions but peace of mind. Our commitment to exceptional customer service means you'll have a dedicated account manager who will be your single point of contact, ensuring a smooth and hassle-free experience."

➢ Overcoming Objections:

"If you have any concerns or questions about how our services work or if we can meet your specific needs, I'm here to address them. We've helped companies of all sizes in various industries overcome similar challenges, and I'm confident we can do the same for you.

➢ Closing:

"To explore how we can work together and provide you with a personalized freight solution, I'd like to schedule a brief consultation. Are you available for a 15-20 minute call sometime this week?"

> **Follow-Up:**

"If now isn't the right time, no problem. I'll follow up with you in a few days to see if you have any further questions or if you're ready to move forward. In the meantime, feel free to visit our website at [Your Website] to learn more about our services."

> **Thank You:**

"Thank you for your time today, [Prospect's Name]. I appreciate the opportunity to discuss how [Your Company Name] can be your trusted logistics partner. I look forward to speaking with you soon."

Feel free to customize this script to fit your specific company's services and the needs of your potential clients. Remember to maintain a friendly and professional tone throughout the conversation.

❖ **Introduction:**

> **Agent:** Good [morning/afternoon], my name is [Your Name], and I represent [Your Company], a trusted and reliable freight brokerage firm. How are you today?

- ➤ **Prospect:** I'm doing well, thank you. How can I help you?

- ➤ **Agent:** I appreciate your time. I'd like to discuss how [Your Company] can streamline your freight logistics and help you save time and money. May I ask a few questions to better understand your needs?

- ➤ **Prospect:** Of course, go ahead.

- ❖ **Building Rapport:**

- ➤ **Agent:** Great. To start, could you briefly describe your current freight logistics process? Are there any pain points or challenges you're facing?

- ➤ **Prospect:** We currently work with multiple carriers, and it's become quite a hassle to manage all of them effectively. Costs are also increasing, and we need more transparency.

- ➤ **Agent:** I understand. Managing multiple carriers can indeed be challenging. [Your Company] specializes in simplifying the freight management process and optimizing costs for businesses like yours.

❖ **Value Proposition:**

➤ **Agent:** Here are a few ways we can benefit your company:

➤ **Streamlined Carrier Selection:** We have an extensive network of vetted carriers, ensuring you get the best service at competitive rates without the hassle of managing multiple relationships.

➤ **Cost Savings:** Our expertise and industry relationships enable us to negotiate better rates on your behalf and identify cost-saving opportunities in your logistics.

➤ **Real-time Tracking:** Our advanced technology provides you with real-time tracking and visibility into your shipments, improving efficiency and reducing delays.

- ➢ **Dedicated Support:** You'll have a dedicated account manager who understands your unique requirements and can provide personalized solutions.

❖ **Handling Objections:**

- ➢ **Prospect:** We've had issues with delays in the past. How can you ensure on-time delivery?

- ➢ **Agent:** That's a valid concern. We take pride in our commitment to on-time delivery. Our advanced tracking systems allow us to monitor shipments in real-time and proactively address any potential issues. Plus, our extensive carrier network means we can quickly find alternatives if needed.

❖ **Closing the Sale:**

- ➢ **Agent:** Given the challenges you've mentioned, I believe [Your Company] can make a significant positive impact on your logistics operations. Would you be open to a more in-depth discussion or a demo of our services? We can tailor a solution to meet your specific needs.

- ➤ **Prospect:** I'm interested. Can you provide more details and pricing information?

- ➤ **Agent:** Absolutely. I'd be happy to provide you with a personalized proposal based on your requirements. Could we schedule a follow-up call, perhaps later this week, to go over the specifics?

- ➤ **Prospect:** Sounds good. Let's set up a call for Thursday.

- ➤ **Agent:** Great! I'll send you an email with some available times for Thursday. Thank you for considering [Your Company], and I look forward to assisting you with your freight logistics needs.

- ➤ **Follow-up:**

After the call, be sure to send a follow-up email with the proposed details and any additional information discussed during the conversation.

Remember, a successful sales script should be tailored to the prospect's needs and challenges. Be prepared to adapt and listen actively during the conversation to address their specific concerns and demonstrate how your freight brokerage services can benefit their business.

❖ **Introduction:**

➢ **Agent:** Good [morning/afternoon], my name is [Your Name], and I represent [Your Company], a trusted and reliable freight brokerage firm. How are you today?

➢ **Prospect:** I'm doing well, thank you. How can I help you?

➢ **Agent:** I appreciate your time. I'd like to discuss how [Your Company] can streamline your freight logistics and help you save time and money. May I ask a few questions to better understand your needs?

➢ **Prospect:** Of course, go ahead.

❖ **Building Rapport:**

➢ **Agent:** Great. To start, could you briefly describe your current freight logistics process? Are there any pain points or challenges you're facing?

- ➢ **Prospect:** We currently work with multiple carriers, and it's become quite a hassle to manage all of them effectively. Costs are also increasing, and we need more transparency.

- ➢ **Agent:** I understand. Managing multiple carriers can indeed be challenging. [Your Company] specializes in simplifying the freight management process and optimizing costs for businesses like yours.

- ❖ **Value Proposition:**

- ➢ **Agent:** Here are a few ways we can benefit your company:

- ➢ **Streamlined Carrier Selection:** We have an extensive network of vetted carriers, ensuring you get the best service at competitive rates without the hassle of managing multiple relationships.
- ➢ **Cost Savings:** Our expertise and industry relationships enable us to negotiate better rates on your behalf and identify cost-saving opportunities in your logistics.

- ➤ **Real-time Tracking:** Our advanced technology provides you with real-time tracking and visibility into your shipments, improving efficiency and reducing delays.

- ➤ **Dedicated Support:** You'll have a dedicated account manager who understands your unique requirements and can provide personalized solutions.

❖ **Handling Objections:**

- ➤ **Prospect:** We've had issues with delays in the past. How can you ensure on-time delivery?

- ➤ **Agent:** That's a valid concern. We take pride in our commitment to on-time delivery. Our advanced tracking systems allow us to monitor shipments in real-time and proactively address any potential issues. Plus, our extensive carrier network means we can quickly find alternatives if needed.

❖ **Closing the Sale:**

- ➤ **Agent:** Given the challenges you've mentioned, I believe [Your Company] can make a significant positive impact on your logistics operations. Would you be open to a more in-depth discussion or a demo of our services? We can tailor a solution to meet your specific needs.

➤ **Prospect:** I'm interested. Can you provide more details and pricing information?

➤ **Agent:** Absolutely. I'd be happy to provide you with a personalized proposal based on your requirements. Could we schedule a follow-up call, perhaps later this week, to go over the specifics?

➤ **Prospect:** Sounds good. Let's set up a call for Thursday.

➤ **Agent:** Great! I'll send you an email with some available times for Thursday. Thank you for considering [Your Company], and I look forward to assisting you with your freight logistics needs.

> **Follow-up:**

After the call, be sure to send a follow-up email with the proposed details and any additional information discussed during the conversation.

Remember, a successful sales script should be tailored to the prospect's needs and challenges. Be prepared to adapt and listen actively during the conversation to address their specific concerns and demonstrate how your freight brokerage services can benefit their business.

Cold email scripts can be an effective way to reach out to potential clients or partners in the freight brokerage industry. Here's a sample cold email script tailored for freight brokers:

> **Subject:** Streamline Your Freight Logistics with [Your Company Name]

Hi [Recipient's Name],

I hope this email finds you well. My name is [Your Name], and I specialize in helping businesses like yours optimize their freight logistics to reduce costs and improve efficiency.

I understand that managing your freight operations can be a complex and time-consuming task. That's where [Your Company Name] can make a significant difference. With [X years/months] of experience in the industry, we have a proven track record of delivering outstanding results.

❖ **Here's how we can assist you:**

➢ **Cost Savings:** We excel in negotiating the best rates with carriers, ensuring you save on shipping costs.

➢ **Efficiency:** Our team utilizes cutting-edge technology to streamline your supply chain, reducing transit times and ensuring on-time deliveries.

➢ **Custom Solutions:** We tailor our services to your specific needs, whether it's full-truckload, LTL, or specialized freight.

➢ **Exceptional Customer Service:** We're committed to providing top-notch customer support, ensuring you have a partner you can rely on.

I'd love the opportunity to discuss how [Your Company Name] can help your business achieve these benefits. Could we schedule a brief call or meeting at your convenience?

Please let me know a date and time that works for you, or if you prefer, you can reply to this email with any questions or specific challenges you're currently facing in your freight logistics.

Thank you for considering us, [Recipient's Name]. We're excited about the possibility of working with [Recipient's Company Name].

Best regards,

[Your Name]

[Your Title]

[Your Company Name]

[Your Phone Number]

[Your Email Address]

[Your Company Website]

This script should serve as a starting point. Be sure to personalize it for each recipient, emphasizing the value you can provide to their specific business needs. Additionally, keep your emails concise, professional, and respectful of the recipient's time.

> **Subject:** Seamless Logistics Solutions for Your Shipping Needs

Dear [Shipper's Name],

I hope this email finds you well. My name is [Your Name], and I represent [Your Company Name], a trusted and experienced freight brokerage firm. We are reaching out to you because we believe we can offer you cost-effective, efficient, and reliable solutions for your shipping needs.

At [Your Company Name], we understand the challenges that shippers like you face in today's dynamic logistics landscape. Whether you're dealing with tight deadlines, complex routes, or fluctuating shipping demands, our team is committed to providing tailor-made solutions that streamline your transportation processes.

❖ **Here's what you can expect when you partner with us:**

Extensive Network: We have established relationships with a vast network of carriers, ensuring you have access to a wide range of transportation options, from LTL (Less-Than-Truckload) to FTL (Full Truckload) and everything in between.

➢ **Cost Savings:** Our industry expertise and negotiation skills enable us to secure competitive rates for your shipments, helping you reduce transportation costs and boost your bottom line.

➢ **Reliability:** We take pride in our commitment to timely deliveries and transparent communication. You can count on us to keep you informed throughout the shipping process.

➢ **Custom Solutions:** We understand that your shipping needs are unique. Our team will work closely with you to develop customized logistics solutions that align with your business goals.

➢ **Cutting-Edge Technology:** We leverage state-of-the-art technology to track, manage, and optimize your shipments, providing you with real-time visibility and control.

We would love the opportunity to discuss your specific requirements and demonstrate how we can add value to your supply chain. Please let us know a convenient time for a brief call or meeting, and we'll be happy to provide you with a personalized proposal.

Thank you for considering [Your Company Name] as your trusted logistics partner. We look forward to the possibility of collaborating and simplifying your shipping operations.

Warm regards,

[Your Name]

[Your Title]

[Your Company Name]

[Your Email]

[Your Phone Number]

Creating a sales script for freight brokers is essential to help you communicate effectively with potential clients and win their business. Below is a sample sales script that you can customize to fit your specific services and target audience:

❖ Introduction:

"Hello [Prospect's Name], my name is [Your Name] from [Your Company Name], and I specialize in freight brokerage services. I hope you're doing well today. I wanted to reach out to discuss how we can help you optimize your shipping and logistics operations. Is this a good time to chat?"

➢ Establish Rapport:

"Before we dive into the details, I'd love to learn more about your business. Can you tell me a bit about your current shipping and logistics needs, as well as any challenges you're facing?"

❖ Highlight Your Expertise:

"At [Your Company Name], we have a wealth of experience in the freight brokerage industry. We've helped countless businesses like yours streamline their supply chain and reduce transportation costs. Our team of experts is well-versed in managing various types of cargo, including [mention specific types or industries you specialize in]."

❖ Address Pain Points:

"I understand that shipping and logistics can be a complex and time-consuming aspect of your business. Many of our clients initially come to us with challenges such as [list common pain points like high shipping costs, unreliable carriers, shipment tracking issues, etc.]. Are any of these concerns familiar to you?"

❖ Solution Presentation:

➢ "Well, the good news is, we can provide a tailored solution to address these challenges. Here's how we can help:

➢ **Carrier Network:** We have a vast network of trusted carriers, allowing us to secure the best rates and ensure on-time deliveries.

➢ **Streamlined Processes:** We simplify the shipping process, from booking to tracking, making it hassle-free for you.

➤ **Cost Savings:** Our expertise in negotiating rates and optimizing routes often leads to significant cost reductions.

➤ **Risk Management:** We ensure that your shipments are adequately insured and comply with all regulations.

➤ **Personalized Service:** You'll have a dedicated account manager who will be available 24/7 to assist you."

➤ **Benefits and ROI:** "Our clients typically see [mention a percentage] reduction in shipping costs and [mention a percentage] increase in overall efficiency within the first [mention a time frame] of partnering with us. We believe we can achieve similar results for your business.

❖ **Call to Action:**

"Would you be open to a more in-depth discussion about how we can specifically tailor our services to meet your needs? We can arrange a free consultation to dive deeper into your logistics requirements."

❖ **Overcome Objections:**

If the prospect raises objections or concerns, be prepared to address them with specific solutions. Common objections include pricing, trust, and competition. Here's an example of how to respond:

➤ **Pricing:** "I understand that cost is a significant concern. However, our track record shows that our services often lead to significant cost savings. We can provide a cost analysis specific to your business to demonstrate the potential return on investment."

➤ **Trust:** "Trust is essential in this industry, and we've built a solid reputation over the years. We can provide references from satisfied clients who can vouch for our reliability and the quality of our services."

➤ **Competition:** "It's always a good practice to compare your options. We encourage you to explore other providers, but we're confident that our combination of experience, service, and cost-effectiveness sets us apart from the competition."

> ➤ **Closing:**

"Let's schedule a call to discuss your specific logistics needs in more detail. I'm confident that we can provide the solutions you're looking for. When is a good time for you to connect?"

Remember to adapt this script to your unique value proposition, services, and industry. The key is to build a genuine connection with your prospects and demonstrate how your freight brokerage services can make their logistics operations more efficient and cost-effective.

Here's a sales script for freight brokers that you can use as a template and customize to fit your specific business and target audience.

Introduction:

> ➤ **Greeting:**

"Hello, [Prospect's Name], this is [Your Name] from [Your Company Name]. How are you today?"

> **Establishing Credibility:**

"I wanted to reach out because I noticed that you're involved in the logistics and transportation industry, and I believe we could be of assistance to your operations."

❖ **Transition:**

> **Identify Pain Points:**

"In this fast-paced industry, we understand that there are numerous challenges, such as fluctuating shipping rates, capacity constraints, and the need for efficient route optimization. Are these issues that you're currently facing?"

> **Show Understanding:**

"We've worked with many companies like yours, and we know how critical it is to keep your supply chain running smoothly while controlling costs."

> **Value Proposition:**

❖ **Highlight Your Services:**

"At [Your Company Name], we specialize in freight brokerage services. We offer a wide range of solutions to address these challenges, including finding reliable carriers, negotiating competitive rates, and ensuring on-time deliveries."

❖ **Benefits of Working with You:**

"By partnering with us, you can focus on your core business while we handle the logistics. Our expertise and industry connections allow us to secure the best rates and ensure your shipments arrive on time, every time."

❖ **Differentiation:**

➢ **Unique Selling Points:**

"What sets us apart from the competition is our commitment to customer service. We're available around the clock to provide real-time updates on your shipments and resolve any issues promptly."

❖ **Call to Action:**

➢ **Ask for a Meeting:**

"I'd love to learn more about your specific needs and discuss how we can tailor our services to benefit your business. Can we schedule a brief meeting or call to explore this further?"

❖ **Overcoming Objections:**

➢ **Address Concerns:**

If the prospect expresses concerns or objections, be prepared to address them with well-thought-out responses. For example, if they mention cost, you could say, "I understand that cost is a significant factor. Our goal is to save you money in the long run by optimizing your logistics and securing competitive rates."

❖ **Closing:**

➢ **Reiterate Benefits:**

"By working together, we believe we can improve your logistics efficiency and reduce your shipping costs. I look forward to the opportunity to discuss this in more detail."

➢ **Express Interest:**

"Is there a convenient time for you to schedule a meeting or call to further explore how our services can benefit your business?"

➢ **Follow-up:**

❖ **If No Immediate Commitment:**

"If you need some time to consider this, that's absolutely fine. I'll follow up with you next week to see if you have any questions or if you're ready to move forward."

➤ **Thank You:**

❖ **Appreciation:**

"Thank you for your time, [Prospect's Name]. I appreciate the opportunity to speak with you today. Have a great day!"

Remember to tailor this script to your specific business, industry, and the unique value you offer to your clients. Additionally, practice your delivery to sound confident, professional, and engaging during your sales calls.

A sales script for freight brokers should be tailored to your specific services and target audience. However, I can provide you with a general template that you can adapt to your needs. Keep in mind that successful sales scripts are often concise, focus on the value you can provide to your customers, and are conversational. Here's a basic template to get you started:

❖ **Introduction:**

➢ **Greeting:** Begin by introducing yourself and your company.

➢ **Example:** "Hello, this is [Your Name] from [Your Company]."

➢ **Purpose:** State the purpose of your call.

➢ **Example:** "I'm reaching out today to discuss how our freight brokerage services can benefit your business."

❖ **Build Rapport:**

➢ **Show Understanding:** Demonstrate that you understand their industry or unique challenges.

➢ **Example:** "I understand that in the freight industry, on-time deliveries and cost-efficiency are critical."

➢ **Ask Questions:** Engage your prospect by asking questions about their current freight operations.

➢ **Example:** "Could you tell me more about how you currently handle your freight shipments?"

❖ **Present Value:**

➢ **Highlight Benefits:** Explain the specific benefits of working with your brokerage.

➢ **Example:** "Our brokerage specializes in finding the most cost-effective and efficient shipping solutions. We can save you time and money."

➢ **Case Studies/Testimonials:** Share success stories or customer testimonials if available.

➢ **Example:** "Many businesses like yours have experienced significant cost savings and improved logistics by partnering with us. Here's an example..."

❖ **Address Objections:**

➢ **Handle Concerns:** Be prepared to address common objections.

➢ **Example:** "I understand you may have concerns about cost. We work closely with a network of carriers to negotiate competitive rates on your behalf."

❖ **Offer a Solution:**

➢ **Customized Approach:** Explain how you tailor solutions to their needs.

➢ **Example:** "We don't believe in one-size-fits-all solutions. We'll work closely with you to develop a freight strategy that fits your unique requirements."

➢ **Next Steps:** Propose the next steps in the sales process.

➢ **Example:** "If this sounds like something that could benefit your business, I'd suggest setting up a meeting to discuss your specific needs and goals."

❖ **Close the Call:**

➢ **Summarize:** Recap the key points discussed.

➢ **Example:** "To summarize, we offer cost-effective freight solutions tailored to your business, and we'd like to explore how we can help you achieve your shipping goals."

➢ **Set Follow-Up:** Arrange a follow-up action.

➢ **Example:** "Would you be available for a meeting next week to discuss this further?"

➢ **Thank You:** Express gratitude for their time.

➢ **Example:** "Thank you for taking the time to speak with me today. I look forward to the possibility of working together."

Remember to customize the script to match your company's unique value proposition and the specific needs of your target customers. Practice and refine your script based on the responses you receive and be ready to adapt it to each prospect's individual situation. Subject: Simplify Your Shipping Process with [Your Company Name]

Dear [Shipper's Name],

I hope this message finds you well. I wanted to reach out to you today because I believe we can help streamline your shipping operations and make your life easier. At [Your Company Name], we understand the challenges that shippers like you face, and we're dedicated to providing efficient and cost-effective solutions.

❖ **Here's how we can make a difference for your business:**

➢ **Tailored Shipping Solutions:** We don't believe in one-size-fits-all solutions. Our team will work closely with you to understand your unique shipping needs and develop a customized plan to optimize your shipping process.

➢ **Cost Savings:** We know that controlling shipping costs is a top priority for you. We have a proven track record of reducing shipping expenses while maintaining service quality, which can directly impact your bottom line.

➢ **Reliability:** We pride ourselves on reliable service. Your shipments will arrive on time, every time, and our customer support team is available 24/7 to assist you with any questions or concerns.

➢ **Advanced Technology:** Our cutting-edge technology platform allows for real-time tracking, reporting, and analytics, giving you full visibility into your shipments and helping you make informed decisions.

➢ **Sustainability:** We're committed to reducing the environmental impact of shipping. We offer eco-friendly options and continuously work on reducing our carbon footprint.

To get started, we'd love to offer you a complimentary consultation to discuss your specific needs and show you how [Your Company Name] can make a positive impact on your shipping operations. We're confident that you'll be pleased with the results.

Please let us know a convenient time for you, and we'll arrange a call or meeting at your convenience. Feel free to reply to this email or call us at [Your Phone Number].

Thank you for considering [Your Company Name] as your trusted shipping partner. We look forward to the opportunity to work together and help you achieve your shipping goals.

Warm regards,

[Your Name]

[Your Title]

[Your Company Name]

[Your Email]

[Your Phone Number]

[Your website]

CHAPTER TWO

WARM SCRIPTS

Warm scripts are indispensable tools for brokers aiming to cultivate relationships and secure business. These scripts, crafted with a balance of professionalism and personalization, serve to maintain communication momentum after initial contact. Expressing gratitude for past interactions, these scripts can tactfully inquire about the client's current needs or offer additional information relevant to their interests.

Effective follow-up scripts also provide a platform for addressing any concerns, reaffirming the broker's commitment to client satisfaction. Incorporating a call to action, such as scheduling a follow-up meeting or property tour, adds purpose to the communication.

In the competitive realm of brokerage, consistent and thoughtful follow-up not only reinforces the broker's presence in the client's mind but also demonstrates a genuine interest in meeting their needs. These scripts, when executed skillfully, contribute to building trust, fostering long-term relationships, and positioning the broker as a reliable partner in the dynamic world of real estate.

Here are some follow-up script templates for different scenarios:

1. Follow-up with a Shipper after Initial Contact:

Subject: Re: [Shipment Name or Reference]

Dear [Shipper's Name],

I hope this message finds you well. I wanted to follow up on our previous conversation regarding your upcoming shipment from [Origin] to [Destination].

Our team is eager to provide you with the best logistics solutions, and we have been diligently working on your request. We've made significant progress and have identified a few viable carrier options for your consideration.

Could we schedule a brief call to discuss the carriers and the next steps? Your input is invaluable in ensuring a smooth and successful transportation process.

Please let me know your availability, and we'll set up a time that works for you.

Thank you for considering [Your Company Name] for your logistics needs. We look forward to helping you achieve your shipping goals.

Warm regards,

[Your Name]

[Your Title]

[Your Company Name]

[Your Phone Number]

[Your Email]

2. **Follow-up with a Carrier after Initial Contact:**

Subject: Re: Carrier Partnership Opportunity

Dear [Carrier's Name],

I trust this message finds you in good health. We appreciate the opportunity to connect and explore the potential for a partnership between [Your Company Name] and [Carrier's Company Name].

After our initial discussion, we were impressed by your company's capabilities and commitment to exceptional service. Our team has been actively reviewing your qualifications, and we believe there could be some great synergies between our organizations.

To move forward, we would like to discuss some specifics, including your available capacity, pricing structure, and any unique services you offer. We believe that aligning our goals will be mutually beneficial.

Could you please let us know when it would be convenient for you to have a more in-depth conversation about this partnership opportunity? We are excited about the possibility of working together.

Thank you for your time and consideration.

Best regards,

[Your Name]

[Your Title]

[Your Company Name]

[Your Phone Number]

[Your Email]

3. Follow-up with an Existing Client:

Subject: Checking in on Your Freight Needs

Dear [Client's Name],

I hope everything is going smoothly for your business. It's been a pleasure working with you at [Your Company Name], and we value your partnership.

I wanted to check in and ensure that our services are still meeting your needs and expectations. Are there any upcoming shipments or any changes in your logistics requirements that we should be aware of? Our goal is to continuously optimize our services for you.

Additionally, if you have any feedback, suggestions, or concerns, please don't hesitate to share them with us. Your input is crucial in helping us provide the best possible service.

Thank you for your ongoing trust in [Your Company Name]. We look forward to continuing to serve your freight logistics needs.

Warm regards,

[Your Name]

[Your Title]

[Your Company Name]

[Your Phone Number]

[Your Email]

These scripts can serve as a starting point, and you can customize them to suit your specific communication style and the needs of your clients, carriers, or shippers. Remember to maintain a professional and friendly tone in your follow-up messages.

Here's a follow-up script for freight brokers to use when reaching out to potential clients or carriers:

[Your Name]: Hello, [Client's/Carrier's Name],

I hope this message finds you well. I wanted to follow up on our recent conversation and see how things are progressing on your end. We're excited about the possibility of working together and would like to discuss the next steps and address any questions or concerns you may have.

Here's what we can offer:

> ➤ **Tailored Solutions:** We understand that every client and carrier have unique needs. Our team is

committed to providing personalized solutions that meet your specific requirements.

> **Competitive Rates:** We work hard to secure the best rates for our clients while ensuring carriers receive fair compensation for their services. Our goal is to create a win-win situation for all parties involved.

> **Reliability:** We pride ourselves on our track record of on-time deliveries and dependable service. You can count on us to keep your supply chain running smoothly.

> **Real-time Tracking:** Our state-of-the-art technology allows you to track shipments in real-time, giving you complete visibility and control over your cargo.

> **Exceptional Customer Service:** Our team is available 24/7 to assist you with any questions or issues. We provide top-notch customer service and support.

Now, I'd like to schedule a call or meeting at your convenience to dive deeper into your specific needs and how we can best serve you. Please let me know a date and time that works for you, or if you prefer to continue the conversation via email or another method, please don't hesitate to let me know.

If there are any specific questions or concerns, you'd like to address, please feel free to share them, and we'll ensure they are addressed promptly.

Thank you for considering us as your freight partner. We're eager to explore how we can collaborate to enhance your logistics and transportation needs. Your satisfaction is our priority.

Looking forward to hearing from you soon.

Best regards,

[Your Name]

[Your Company]

[Your Contact Information]

Feel free to customize this script with specific details about your services and any relevant information about your previous conversations with the client or carrier. It's important to be genuine, helpful, and open to addressing their individual needs and concerns.

A follow-up script for freight brokers is an essential tool to maintain and strengthen relationships with carriers and shippers. Effective communication is key in the logistics industry, and follow-ups demonstrate your commitment to providing excellent service. Here's a sample follow-up script for freight brokers to use as a starting point:

Subject: Follow-Up on Recent Shipment and Future Opportunities

Dear [Carrier's or Shipper's Name],

I hope this message finds you well. I wanted to take a moment to follow up on our recent collaboration and discuss potential opportunities for future partnerships. At [Your Brokerage Company Name], we greatly value the relationships we've built with our partners, and we are committed to delivering the best service to our clients.

1. Gratitude and Acknowledgment:

First and foremost, I want to express our appreciation for your support and hard work on the recent shipment [include shipment ID or reference]. Your dedication and professionalism were crucial in ensuring the successful delivery, and we received positive feedback from our client.

2. Feedback Request:

We believe in continuous improvement, and your feedback is important to us. Is there anything you believe we could have done differently to make the process smoother for you? Your insights are invaluable in helping us enhance our service quality.

3. Future Opportunities:

We're always on the lookout for new projects and partnerships. If you have upcoming capacity or specific lanes you're interested in, please let us know. We're eager to explore how we can align your capabilities with our clients' needs.

4. Competitive Rates and Terms:

Our goal is to ensure that our partners receive competitive rates and favorable terms. If you have any specific rate or contract requests, don't hesitate to reach out. We're here to negotiate and work with you to find a mutually beneficial arrangement.

5. Commitment to Communication:

Clear and open communication is the foundation of any successful partnership. Please know that we're here for you whenever you need us, whether it's to discuss upcoming opportunities, address concerns, or just touch base.

6. Upcoming Events or Updates:

We'd like to keep you informed about industry events, regulatory changes, or any updates that may impact your business. Please stay tuned for our newsletters and announcements.

7. Next Steps:

Should you be interested in collaborating on future shipments or have any questions, please feel free to contact us at [Your Contact Information].

Closing:

Thank you again for your partnership and trust in [Your Brokerage Company Name]. We look forward to working together on exciting opportunities in the near future.

Warm regards,

[Your Name]

[Your Title]

[Your Brokerage Company Name]

[Your Contact Information]

Remember to personalize this script with the recipient's name, details of your recent collaboration, and any specific information relevant to your ongoing relationship. The goal is to create a sense of partnership, reliability, and open communication, fostering a strong and lasting business relationship.

Subject: Follow-up on Shipment Status

Dear [Shipper's Name],

I hope this email finds you well. I wanted to provide you with a follow-up on the status of the shipment we discussed earlier. As your dedicated freight broker, I am committed to ensuring the smooth transportation of your goods, and I understand the importance of keeping you informed throughout the process.

As of our last communication, the shipment was scheduled for pick-up on [Date] from your location. Our carrier, [Carrier Name], has been assigned to this shipment and is ready to execute the transportation according to our agreement.

Our team will be closely monitoring the progress of this shipment, and I will be your point of contact for any updates or changes that may arise during transit. If you have any specific instructions, additional requirements, or concerns related to this shipment, please don't hesitate to reach out to me at [Your Contact Information].

I understand that your cargo is valuable, and I want to assure you that we are dedicated to its safe and timely delivery. Our commitment to providing top-quality service is unwavering, and I am here to assist you at every step of the way.

Thank you for entrusting us with your freight needs, and please feel free to contact me if you have any questions or require further information. We greatly appreciate your business and look forward to continuing our successful partnership.

Best regards,

[Your Name]

[Your Title]

[Your Company]

[Your Contact Information]

CHAPTER THREE

NEGOTIATING SCRIPTS

Negotiating scripts is a pivotal skill for brokers, as it directly impacts their earnings and client relationships. Brokers must adeptly balance their desire for competitive compensation with the need to provide value to clients. Successful rate negotiations involve a deep understanding of market trends, property values, and the specific needs of clients. Brokers often employ a consultative approach, thoroughly explaining their expertise, the intricacies of the market, and the benefits of their services. Transparency about costs and a willingness to tailor packages to client requirements enhance the negotiation process. Establishing a rapport and aligning interests with clients fosters a collaborative atmosphere, increasing the likelihood of reaching a mutually beneficial agreement. In the dynamic world of brokerage, the ability to navigate rate negotiations with finesse not only secures fair compensation for brokers but also solidifies trust and credibility, laying the groundwork for enduring client partnerships.

Negotiation is a critical skill for freight brokers, as it can have a significant impact on your success in the transportation industry. Here are some negotiation techniques and tips for freight brokers:

❖ **Understand Your Market:**
➢ Stay informed about market conditions, including supply and demand, pricing trends, and capacity constraints. This knowledge will give you a better understanding of what is negotiable and what isn't.
➢ Build Strong Relationships:
➢ Develop and maintain strong relationships with shippers, carriers, and other industry stakeholders. Trust and rapport can make negotiations smoother.

❖ **Set Clear Objectives:**
➢ Before entering into a negotiation, define your objectives and desired outcomes. Be specific about the rates, terms, and conditions you're aiming for.

❖ **Gather Information:**
➢ Collect as much information as possible about the shipment, including the cargo's specifics, the desired delivery time, and any special requirements. The more you know, the better you can negotiate.

❖ **Listen Actively:**

➢ Pay close attention to what the other party is saying. Listen for their needs and concerns and use this information to tailor your offers accordingly.

❖ **Highlight Your Value:**

➢ Emphasize what makes your services unique and why the other party should choose you over your competitors. This could be your reliability, speed, or industry expertise.

❖ **Be Patient:**

➢ Negotiations can take time. Don't rush the process and be willing to give the other party time to consider your offers.

❖ **Negotiate Win-Win Deals:**

➢ Strive for mutually beneficial agreements. A win-win approach is more likely to lead to long-term, profitable relationships.

❖ **Use Silence:**

➢ Sometimes, silence can be a powerful tool in negotiation. After making an offer, give the other party a moment to respond. They might reveal their position or make a counteroffer.

❖ Offer Multiple Options:

➢ Present more than one option to the other party. This gives them a sense of control and can lead to a more favorable agreement for both sides.

❖ Know Your Limits:

➢ Establish your bottom line and be prepared to walk away if the terms are not favorable. Knowing your limits can prevent you from making unfavorable deals out of desperation.

❖ Use Technology:

➢ Leverage technology and data to analyze and optimize your pricing and routing decisions. Software and data analytics can provide valuable insights.

❖ Be Flexible:

➢ Be open to compromise and adapt to changing market conditions. Flexibility can help you find common ground in negotiations.

❖ **Negotiate Long-Term Contracts:**

➢ If possible, negotiate long-term contracts with shippers and carriers. These agreements can provide stability and better rates over time.

❖ **Document Agreements:**

➢ Once an agreement is reached, document it in a clear and concise contract. This helps prevent misunderstandings and disputes in the future.

❖ **Learn from Experience:**

➢ Reflect on your negotiations, especially after challenging ones. Identify what worked and what didn't and use this knowledge to improve your future negotiations.

➢ Remember that negotiation is an ongoing skill that you can continuously develop. By honing your negotiation techniques and strategies, you can enhance your effectiveness as a freight broker and build successful, long-lasting business relationships.

Here's a script that can help guide your rate negotiation with carriers or shippers. Keep in mind that effective negotiation also involves active listening and adapting to the specific situation. This script is just a starting point:

Introduction:

"Hello [Carrier/Shipper's Name], I hope you're doing well. My name is [Your Name] with [Your Company Name]. We're interested in working with you on a new shipment, and we'd like to discuss the rates. Before we get into the details, can you briefly tell me about your expectations and any specific requirements you have for this shipment?"

❖ **Establishing Rapport:**

➢ Show genuine interest in their business.

➢ Mention any previous successful collaborations or shared experiences if applicable.

❖ **Understanding Their Needs:**

➢ "I want to make sure we meet your needs and expectations. Could you tell me more about the specific requirements for this shipment? Are there any time constraints or unique challenges we should be aware of?"

❖ **Presenting Your Offer:**

➢ "I understand your requirements better now. Based on our analysis, we can offer you [Your Initial Rate] for this shipment. This rate considers factors such as distance, shipment size, and any additional services you might require. We believe this is a competitive offer, but we're open to discussing and adjusting it if necessary."

❖ **Handling Objections:**

➢ If they find your rate too high: "I understand your concern about the rate. Can you help me understand your budget constraints? Perhaps we can find a compromise that aligns with both our goals."

➢ If they mention a lower competing offer: "I appreciate your honesty. Our priority is to provide quality service. Are there any specific services or value-added features we offer that could justify the slightly higher rate?"

➢ If they ask for a discount: "We appreciate your business and are willing to work with you. However, I'd like to understand your reasons for seeking a discount. If we can find ways to streamline the process or offer additional services, would that help justify the rate?"

❖ **Closing the Deal:**

➢ "I'm committed to finding a solution that works for both of us. Let's explore some possible compromises and additional services that could make this partnership beneficial. How about we [e.g., discuss a longer-term contract, explore ways to optimize the route, or offer additional tracking services]? Does that sound like a step in the right direction?"

❖ **Summarizing the Agreement:**

➢ "Great, I believe we're making progress. To summarize, we're looking at [Final Agreed Rate] for this shipment, and we'll also provide [additional services] to ensure a smooth and efficient process. Does this align with your expectations?"

❖ **Confirmation and Next Steps:**

➢ "Fantastic. I appreciate your willingness to work with us on this. I'll send you a written agreement with all the details we discussed, and we can move forward from there. If you have any questions or need clarification, feel free to reach out anytime."

➢ Remember, successful negotiation isn't just about getting the lowest rate but creating a win-win situation. Be flexible and open to compromise while ensuring that the deal is profitable for both parties.

Here's a script you can use as a starting point for negotiating rates with shippers or carriers as a freight broker. Remember that effective negotiation often involves flexibility, active listening, and adapting to the specific situation. Feel free to tailor this script to your needs and the specific circumstances of your negotiation.

Introduction:

"Hello [Shipper/Carrier's Name],

I hope you're doing well. My name is [Your Name], and I'm a freight broker with [Your Company Name]. I've been reviewing the details of the shipment you're looking to move, and I'm excited to discuss rates with you to ensure we can provide the best service.

To begin, could you please provide me with some essential information about this shipment, such as the origin, destination, type of cargo, weight, and any special requirements?"

❖ **Gathering Initial Information:**

➤ Ensure you have all necessary shipment details.

➤ Show genuine interest in the shipper/carrier's needs and concerns.

❖ **Present Your Understanding:**

➤ "Thank you for sharing those details. It seems like a [briefly summarize the shipment]. Based on what you've provided, I understand that [mention any special considerations or requests]. Is there anything else I should know before we proceed with rate negotiation?"

❖ **Acknowledging Value:**

➤ Show that you respect their time and expertise.

➤ Build rapport and demonstrate understanding.

❖ **Rate Negotiation:**

➤ "Great, let's move on to discussing rates. Based on the information you've given me and the current market conditions, I would suggest a starting point of [your initial rate offer] for this shipment. I believe this is a fair rate because [briefly justify your rate based on market conditions, service quality, or other factors]."

❖ **Listen and Encourage Feedback:**
➢ Give the other party an opportunity to respond.
➢ Be open to their feedback and concerns.

❖ **Handling Objections:**
➢ "I understand that you might have some concerns or questions about the rate. Could you please share your thoughts or any specific issues you see with the proposed rate?"

❖ **Counter-Offer:**
➢ If they raise concerns, address them professionally.
➢ Offer a counter-offer that you're comfortable with but is still favorable to both parties.
➢ "I appreciate your feedback, and I want to find a rate that works for both of us. Given your concerns, I'd be willing to adjust the rate to [your counter-offer]. This rate reflects the current market conditions and the value we provide in terms of [mention benefits of your services]. Does this rate work better for you, or is there anything else we should consider?"

❖ **Finalizing the Agreement:**

➤ Once you've reached an agreement, summarize the terms.

➤ Ask for confirmation to ensure both parties are aligned.

➤ "So, just to confirm, we've agreed on a rate of [final rate]. I'll send over the contract or confirmation for your review. Is there anything else you'd like to discuss before we proceed with the paperwork?"

❖ **Closing:**

➤ Express gratitude for the negotiation process.

➤ Reiterate your commitment to providing excellent service.

➤ "Thank you for your time and cooperation during this negotiation. We're looking forward to working with you and ensuring a smooth and successful shipment. If you have any further questions or need assistance, please don't hesitate to reach out. Have a great day!"

This script is a guideline, and you should adapt it to the specific situation and parties involved. Remember that successful negotiations often involve compromise, active listening, and a focus on building a mutually beneficial relationship.

CHAPTER FOUR

HOW TO PROSPECT SHIPPERS

Prospecting shippers is a crucial part of a freight broker's job.

Here are steps to help you effectively prospect shippers as a freight broker:

- ❖ **Identify Your Target Market:**
  - ➤ Start by defining your target market. What types of shippers are you looking for? Consider factors like the industries they operate in, their location, and the type of freight they typically need to move.

- ❖ **Research Potential Shippers:**
  - ➤ Use various resources to research potential shippers. This can include industry directories, online databases, trade associations, and business directories. You can also search for companies in your target market on the internet.

- ❖ **Build a Prospecting List:**
  - ➤ Create a list of potential shippers based on your research. Include their contact information, such as

company name, address, phone number, and email address. Tools like CRM software can be very helpful for managing your prospecting list.

❖ **Develop a Value Proposition:**

➢ Before reaching out to shippers, understand what sets you apart from the competition. What value can you offer to shippers? This could include competitive pricing, excellent service, access to a vast carrier network, or specialized knowledge in their industry.

❖ **Cold Calling and Email Outreach:**

➢ Start reaching out to your prospects via phone calls and emails. Craft a compelling and concise message that highlights the value you can provide. Be prepared to introduce yourself, your company, and explain how you can meet their shipping needs.

❖ **Networking and Industry Events:**

➢ Attend industry events, conferences, and trade shows to network with potential shippers. These events provide an excellent opportunity to make direct connections and establish relationships.

❖ **Utilize social media and Online Marketing:**

➢ Establish a strong online presence through social media platforms and a professional website. Share informative content and engage with potential shippers on platforms like LinkedIn.

❖ **Ask for Referrals:**

➢ Leverage your existing relationships with carriers, customers, and partners to ask for referrals to potential shippers. Personal recommendations can carry significant weight in the industry.

❖ **Follow Up:**

➢ Don't be discouraged by initial rejections or non-responses. Persistence is key in prospecting. Follow up with prospects who haven't responded and maintain regular contact with those who have shown interest.

❖ **Provide Exceptional Customer Service:**

➢ Once you've secured a shipper as a customer, provide exceptional service to build a strong and lasting relationship. Happy customers can lead to repeat business and referrals.

❖ **Stay Informed:**

➢ Keep yourself informed about the shipping industry, market trends, and regulatory changes. This knowledge can help you better serve your customers and differentiate yourself from the competition.

❖ **Measure and Adjust:**

➢ Track your prospecting efforts and measure what is working and what isn't. Adjust your strategy accordingly to improve your results over time.

Remember that successful prospecting often requires a combination of strategies, and it may take time to see significant results. Building a solid reputation and relationships within the industry can be one of the most valuable assets for a freight broker.

CHAPTER FIVE

WHERE TO FIND SHIPPERS

It's important to note that finding shippers involves a combination of research, networking, and persistence. Finding shippers' email addresses can be a challenging task, as many businesses keep this information private for various reasons, including privacy and security concerns. It's important to note that attempting to obtain someone's email address without their consent can be considered unethical or even illegal in some cases, depending on local laws and regulations. Therefore, it's essential to respect privacy and follow ethical guidelines when attempting to contact potential shippers.

Freight brokers should conduct due diligence on potential shippers to ensure they meet the necessary legal and financial requirements, and they should also be prepared to negotiate terms and rates that benefit both parties.

Freight brokers can find shippers through a variety of methods and resources. Building a strong network and utilizing different strategies can help brokers connect with potential shippers.

Here are some common ways to find shippers:

❖ **Trade Shows and Events:** Attend industry events, conferences, and trade shows related to logistics and transportation. These events provide opportunities to network with shippers and learn about their needs. You can meet potential shippers in person. Collect business cards or contact information during these events.

❖ **Company Websites:** Visit the websites of potential shippers. Sometimes, companies list contact information, including email addresses, on their websites. Look for "Contact Us" or "About Us" pages. Visiting the websites of manufacturers, distributors, and companies that regularly ship goods can be an effective way to identify potential shippers. Many companies list contact information or instructions for carriers and brokers interested in doing business with them.

❖ **Local Business Directories:** Local business directories or chamber of commerce websites can be a valuable resource for finding shippers in a specific geographic area. Contacting local businesses directly may yield opportunities for collaboration.

- ❖ **Third-Party Tools and Services:** There are tools and services available that can help you find email addresses by searching through public data sources. Some examples include Hunter.io, ZoomInfo, and Clearbit.

- ❖ **Online Directories:** There are online directories and databases that compile information about shippers. Examples include ZoomInfo, Kompass, and Manta. These directories may require a subscription or access fee.

- ❖ **Cold Calling and Emailing:** Outreach to potential shippers directly through phone calls and emails. Research companies that may require transportation services and introduce your brokerage's capabilities.

- ❖ **Referrals:** Ask existing shippers and carriers for referrals. They may know other companies in need of transportation services and can connect you with them.

- ❖ **Professional Networking:** Build a strong network within the transportation and logistics industry. Join industry associations, participate in forums, and connect with potential shippers through social media platforms. Join industry-related forums, online communities, or professional networks. Engaging in discussions and networking with professionals in your industry might lead to email exchanges.

❖ **LinkedIn:** LinkedIn can be a valuable resource for finding business contacts. Connect with individuals who work for potential shipping companies, and if they accept your connection request, you may be able to see their email address on their profile.

❖ **Online Marketing:** Create a professional website and utilize online marketing strategies such as search engine optimization (SEO) and social media advertising to attract shippers to your brokerage.

❖ **Direct Mail:** Send targeted direct mail campaigns to potential shippers in your region or industry niche.

❖ **Industry Associations:** Freight broker associations and industry groups often provide directories and resources to connect brokers with potential shippers. Examples include the Transportation Intermediaries Association (TIA) and the National Customs Brokers & Forwarders Association of America (NCBFAA).

❖ **Establish Relationships:** Build and maintain strong relationships with shippers you work with. Satisfied customers are more likely to refer you to others.

❖ **Freight Broker Software:** Invest in freight broker software that can help you identify potential shippers and manage your operations efficiently. Some software solutions include customer relationship management (CRM) features. Other freight brokerage software solutions offer tools to help brokers find and manage shipper information, such as Ascend TMS and McLeod Software.

- ❖ **Subscription Services:** Some companies offer subscription-based services that provide detailed information about potential shippers, including their financial stability and contact details. Examples include Redwood, PIERS, and Panjiva.

- ❖ **Industry Publications:** Trade magazines and publications related to the transportation and logistics industry may include advertisements from shippers seeking transportation services. Subscribing to or browsing these publications can be helpful.

- ❖ **Industry Directories:** Access industry directories and databases that list potential shippers by industry, location, or shipping needs.

- ❖ **Freight Forwarders and 3PLs:** Collaborate with freight forwarders and third-party logistics providers (3PLs) who may have connections with shippers seeking transportation services.

❖ **Local Chambers of Commerce:** Join local chambers of commerce or business associations to connect with businesses in your area that may require transportation services.

Remember that you should always respect privacy and follow legal and ethical guidelines when collecting and using shipper information. Building reputation for reliability, efficiency, and excellent customer service is crucial for attracting and retaining shippers.

And always remember to respect privacy laws and regulations, when collecting and using email addresses. It's crucial to obtain consent for any email communications and ensure that you are compliant with relevant data protection laws in your jurisdiction.

Additionally, building a professional reputation in your industry and maintaining ethical business practices can lead to more open and willing communication with potential shippers, making it easier to obtain their contact information.

CHAPTER SIX

WEBSITE

A well-crafted website is an indispensable asset for brokers, serving as a virtual storefront and a powerful tool for client engagement. A broker's website should be visually appealing, intuitive to navigate, and rich in informative content. Essential features include listings, market insights, and a clear display of the broker's expertise and services.

An effective website not only showcases available properties but also communicates the broker's unique value proposition. Integration of user-friendly contact forms and communication channels facilitates seamless interaction between brokers and potential clients. Mobile responsiveness is crucial, given the increasing prevalence of property searches on smartphones.

Moreover, brokers can leverage their websites for thought leadership by regularly updating content, such as blog posts or market reports. Social media integration further expands the online reach. In an era where the digital presence is often the first point of contact, a well-designed and informative website is an invaluable tool for brokers to establish credibility and attract clients.

To make a website, you'll need to follow these general steps:

❖ **Define Your Purpose and Goals:**
➢ Determine the purpose of your website (e.g., personal blog, business website, portfolio).
➢ Set clear goals for what you want to achieve with your website.

❖ **Choose a Domain Name:**
➢ Select a unique and relevant domain name for your website. You can purchase domain names from domain registrars like GoDaddy, Namecheap, or Google Domains.

❖ **Select a Hosting Provider:**
➢ Choose a web hosting provider to store your website's files and make it accessible on the internet. Popular options include Bluehost, HostGator, and Site Ground.

❖ **Select a Website Building Platform:**
➢ Choose a website building platform or content management system (CMS). Some popular options

are WordPress, Wix, Squarespace, and Shopify (for e-commerce).

❖ **Design Your Website:**

➢ Customize the website's appearance by selecting a theme or template that suits your style and purpose. Add your logo, branding elements, and customize the layout and colors.

❖ **Create and Organize Content:**

➢ Write and organize your website's content, including pages like the homepage, about, services, and contact information.

➢ Add images, videos, and other media as needed.

❖ **Optimize for SEO:**

➢ Optimize your website for search engines (SEO) by using relevant keywords in your content, optimizing images, and creating meta tags.

➢ Install an SEO plugin if you're using a CMS like WordPress.

❖ **Test and Review:**

➢ Thoroughly test your website on different devices and browsers to ensure it looks and functions correctly.

➢ Review your content for accuracy and clarity.

❖ **Add Essential Features:**

➢ Depending on your website's purpose, you may need to add features such as contact forms, social media integration, e-commerce functionality, or a blog.

❖ **Security and Backups:**

➢ Implement security measures to protect your website from threats like malware and hacking.

➢ Set up regular backups to ensure you can restore your site in case of data loss.

❖ **Launch Your Website**:

➢ Once you're satisfied with your website, it's time to publish it. Make it live for the public to see.

❖ **Promote Your Website:**

➢ Share your website on social media, include it in your email signature, and use other marketing strategies to attract visitors.

❖ **Monitor and Maintain:**

➢ Regularly update your content to keep it fresh and relevant.

➢ Monitor website analytics to track visitor behavior and make improvements as needed.

❖ **Seek Help When Needed:**

➢ If you encounter technical difficulties or need advanced features, consider hiring a web developer or designer.

Remember that creating a website can range from a simple one-page site to a complex e-commerce platform, so the specific steps and tools you use may vary based on your goals and technical expertise.

CHAPTER SEVEN

LINKEDIN

LinkedIn is an indispensable platform for brokers, serving as a dynamic hub for professional networking, client engagement, and industry visibility. A well-optimized LinkedIn profile for brokers should showcase their expertise, achievements, and unique value proposition. Regularly sharing industry insights, market trends, and success stories not only establishes credibility but also keeps the broker top-of-mind among their network.

The platform's advanced search features enable brokers to identify potential clients, collaborators, and industry influencers. Engaging in relevant groups and participating in discussions enhances visibility and positions the broker as an authority in their field. Additionally, leveraging LinkedIn's publishing platform to share thought leadership articles or property highlights can further amplify the broker's online presence.

LinkedIn is not only a tool for connecting with existing clients but also a valuable resource for prospecting and staying abreast of industry developments. Brokers who strategically utilize LinkedIn often find it to be a powerful ally in expanding their professional reach and cultivating a robust real estate network.

Here are some steps to help freight brokers find shippers on LinkedIn:

❖ **Optimize Your LinkedIn Profile:**
  ➢ Ensure your LinkedIn profile is complete and professional. Use a clear profile picture, write a compelling headline, and provide a detailed summary of your experience as a freight broker. Highlight your expertise in logistics and transportation.

❖ **Define Your Target Audience:**
  ➢ Determine the specific types of shippers you are looking for. Consider factors like industry, location, company size, and shipping volume. Having a clear target audience in mind will help you refine your search.

❖ **Use Advanced Search:**
  ➢ Utilize LinkedIn's advanced search feature to find potential shippers. You can filter results

based on keywords, industry, location, company size, and more. Use relevant keywords such as "shipping manager," "logistics coordinator," or "supply chain manager" to narrow down your search.

❖ **Join Relevant Groups:**

➢ Join LinkedIn groups related to logistics, transportation, and the industries you are targeting. Participate in discussions and engage with group members. This can help you establish credibility and make connections within your target market.

❖ **Connect Strategically:**

➢ Send personalized connection requests to potential shippers. In your connection request, explain your role as a freight broker and how you can add value to their shipping needs. Avoid generic or spammy messages.

❖ **Engage with Content:**

➢ Stay active on LinkedIn by sharing relevant industry news, insights, and updates. Engage with the content of your connections, especially those in your target industry. Commenting on their posts can help you start conversations.

❖ **Direct Messaging:**
  ➢ Once you've connected with potential shippers, send them a direct message to introduce yourself and your services. Be concise, professional, and focused on how you can help solve their shipping challenges.

❖ **Attend LinkedIn Events:**
  ➢ LinkedIn often hosts virtual events and webinars related to various industries. Participate in these events to network with potential shippers and learn more about their needs and challenges.

❖ **Provide Value:**
  ➢ Offer valuable information, resources, or insights to your connections. This can help build trust and credibility, making potential shippers more likely to consider your services.

- ❖ **Monitor and Follow Up:**
  - ➢ Keep track of your connections and conversations. Follow up with leads to nurture relationships and convert them into clients.

Remember that building meaningful relationships on LinkedIn takes time and effort. It's essential to be genuine and focused on providing value to your connections rather than being overly sales-oriented. Over time, you can develop a strong network of shippers who trust you to handle their freight needs.

Finding shippers on LinkedIn involves using specific search techniques and keywords to locate professionals or companies in the shipping industry.

Here's a step-by-step guide on how to do it:

- ❖ **Create or log in to Your LinkedIn Account:**
  - ➢ If you don't have a LinkedIn account, sign up for one. If you already have an account, log in.

- ❖ **Complete Your Profile:**
  - ➢ Before searching for shippers, make sure your LinkedIn profile is complete and looks professional. A complete profile increases your chances of connecting with others.

❖ **Use Relevant Keywords:**

➢ Use keywords related to shipping, logistics, and the specific type of shipper you are looking for. Examples of relevant keywords include "freight shipping," "logistics manager," "cargo shipping," "import/export manager," etc.

❖ **Use Boolean Search Operators:**

➢ To narrow down your search, use Boolean operators like "AND," "OR," and "NOT" to combine or exclude keywords. For example, you can search for "freight AND shipping NOT courier" to filter out courier companies.

❖ **Use Filters:**

➢ LinkedIn offers various filters to refine your search. You can filter by location, industry, company size, job title, and more. These filters can help you find shippers in your specific target market.

❖ **Browse Company Pages:**

➢ You can also visit the LinkedIn pages of companies that are known to be involved in shipping and logistics. From there, you can often find employees and decision-makers within those companies.

- ❖ **Connect and Network:**
  - ➤ When you find relevant professionals or companies, send them connection requests with a personalized message explaining your intent. Building a professional network on LinkedIn can lead to valuable connections.

- ❖ **Join Relevant Groups:**
  - ➤ LinkedIn has groups dedicated to various industries, including shipping and logistics. Joining these groups can help you connect with professionals in the field and learn from their discussions.

- ❖ **Engage and Participate:**
  - ➤ Actively engage with the content posted by professionals in the shipping industry. Comment on their posts, share valuable insights, and establish yourself as a knowledgeable and active member of the community.

- ❖ **Use LinkedIn Premium (Optional):**
  - ➢ If you have a LinkedIn Premium subscription, you may have access to additional search filters and tools that can help you find shippers more efficiently.

- ❖ **Attend Industry Events and Webinars:**
- ➢ Keep an eye out for shipping industry events, webinars, and conferences on LinkedIn. Participating in these events can help you connect with shippers and stay updated on industry trends.
- ➢ Remember that LinkedIn is a professional networking platform, so it's essential to approach connections and interactions in a professional and respectful manner. Building relationships takes time, so be patient and focused on your search for shippers.
- ➢ Connecting with shippers on LinkedIn can be a valuable way to expand your network and explore potential business opportunities in the shipping and logistics industry. Here's a step-by-step guide to help you connect with shippers on LinkedIn:

❖ **Optimize Your LinkedIn Profile:**

> ➤ Before reaching out to potential shippers, make sure your LinkedIn profile is professional and well-optimized. This includes a clear profile picture, a compelling headline, a summary that highlights your expertise, and a detailed work history.

❖ **Define Your Objectives:**

> ➤ Determine your specific goals for connecting with shippers. Are you looking to collaborate on projects, explore job opportunities, or simply expand your professional network? Knowing your objectives will guide your approach.

❖ **Research Your Target Audience:**

> ➤ Use LinkedIn's search function to find shippers within your industry. You can search by keywords, job titles, and company names. Research their profiles to understand their roles, backgrounds, and any mutual connections.

❖ **Personalize Connection Requests:**
  ➤ When sending connection requests, always include a personalized message. Mention why you want to connect and how the connection could be mutually beneficial. This shows that you've taken the time to research and are genuinely interested.

❖ **Engage in Group Discussions:**
  ➤ Join LinkedIn groups related to shipping, logistics, and related industries. Participate in group discussions, share your insights, and connect with members who are active and relevant to your goals.

❖ **Share Relevant Content:**
  ➤ Regularly post and share content related to shipping and logistics. This can be industry news, your own insights, or articles from reputable sources. Engaging content will help you gain visibility and credibility.

❖ **Comment and Engage:**

➢ Engage with the content shared by shippers and other industry professionals. Leave thoughtful comments and start conversations. Meaningful interactions can lead to connection requests and more significant discussions.

❖ **Attend Webinars and Virtual Events:**

➢ Many professionals in the shipping industry participate in webinars and virtual events. Attend these events, ask questions, and engage with speakers and attendees. You can often connect with participants directly through the event platform or later on LinkedIn.

❖ **Follow Company Pages:**

➢ Follow the LinkedIn company pages of shipping and logistics companies. This will help you stay updated on their activities, job openings, and industry trends.

❖ **Be Patient and Courteous:**

➢ Building meaningful connections on LinkedIn takes time. Be patient and respectful of others' time and boundaries. If someone doesn't accept your connection request or respond to your message, don't be discouraged. Keep building your network.

❖ **Follow Up Professionally:**
  ➢ After connecting with shippers, follow up professionally. You can send a thank-you message for connecting or, if you've discussed potential collaborations, inquire about progress. Maintain a courteous and respectful tone.

❖ **Consider Using Premium Features:**
  ➢ LinkedIn offers premium features that allow you to send InMail messages to people you're not connected with. These can be useful for reaching out to shippers outside your immediate network.

❖ **Respect Privacy and Guidelines:**
  ➢ Always respect LinkedIn's terms of service and privacy guidelines. Avoid spamming or sending unsolicited messages. Build your network organically and professionally.
  ➢ Connecting with shippers on LinkedIn can open up various opportunities in the shipping and logistics

industry. Remember that building a meaningful network takes time and effort, so be persistent, and focus on building authentic and mutually beneficial relationships.

CHAPTER EIGHT

CALL TO ACTION

A "Call to Action" (CTA) is a specific prompt or message designed to encourage the audience or the recipient to take a particular action. CTAs are often used in various forms of marketing materials, websites, emails, advertisements, and more to guide people toward actions such as making a purchase, signing up for a newsletter, clicking a link, or any other desired behavior.

CTAs are indeed an essential part of many marketing and communication strategies. They are used to engage the audience and drive them to take action, which aligns with the broader goals of the strategy. The effectiveness of a CTA can greatly impact the success of a marketing or communication campaign. A "Call to Action" (CTA) is a crucial element in marketing and sales strategies. It's a clear and concise directive that prompts your audience to take specific action. Whether you're creating content for a website, email, social media, or any other marketing channel, a well-crafted CTA can significantly impact your conversion rates and drive desired outcomes.

Here's some information about CTAs and how to use them effectively:

- ❖ **Key Components of a Call to Action:**

  - ➢ **Clarity:** Your CTA should be clear and easy to understand. Use concise language that leaves no room for ambiguity.

  - ➢ **Action-Oriented Verbs:** Begin your CTA with a strong, action-oriented verb that tells the audience what to do. Common action verbs include "buy," "subscribe," "download," "register," "learn," and "contact."

  - ➢ **Benefits:** Communicate the benefits or value the audience will receive by taking the action. Explain why it's in their best interest to act.

  - ➢ **Placement:** The placement of your CTA matters. It should be prominently displayed where it's easily noticeable, such as at the end of a blog post, in a prominent box on a webpage, or within an email's content.

  - ➢ **Design:** The visual design of your CTA can impact its effectiveness. Use contrasting colors, bold fonts, and buttons to make it stand out.

➤ **Mobile Optimization:** Ensure your CTA is mobile-friendly. With the increasing use of smartphones, many interactions occur on mobile devices. Make sure your CTA is easy to tap or click on a small screen.

❖ **Examples of Effective CTAs:**

➤ **"Buy Now":** A classic CTA for e-commerce websites. It's direct and encourages immediate purchasing.

➤ **"Subscribe for Exclusive Updates":** This CTA appeals to those who want to stay informed and receive special content or offers.

➤ **"Download Your Free eBook":** Offers something valuable (an eBook) in exchange for an action (downloading).

➤ **"Request a Quote":** Appropriate for service-based businesses, this CTA encourages potential clients to take the first step in the sales process.

➤ **"Learn More":** Useful when you want to provide additional information without overwhelming the reader.

➢ "Get Started Today": Motivates the audience to begin a process or sign up for a service.

❖ **Testing and Optimization:**

➢ It's important to A/B test different CTAs to see which ones perform best. This involves creating variations of your CTAs and measuring which one generates the most conversions. Factors to test might include wording, color, size, placement, and the use of accompanying visuals.

❖ **CTAs in Different Marketing Channels:**

➢ **Websites:** Place CTAs strategically on your website's pages, especially the homepage, product pages, and landing pages.

➢ **Emails:** In email marketing, the CTA often appears as a clickable button or link, driving recipients to a landing page or taking specific actions.

➢ **Social Media:** Include CTAs in your social media posts, encouraging followers to engage with your content or visit your website.

➢ **Content Marketing:** Blogs, articles, and videos can benefit from CTAs that guide readers/viewers to related content or subscription options.

In conclusion, a well-crafted Call to Action is a vital tool in marketing and sales. It guides your audience to take the desired actions, helping you achieve your goals, whether it's making a sale, gaining a subscriber, or driving engagement. By following the principles of clarity, action-oriented language, and audience benefit, you can create compelling CTAs that drive results.

In a broader strategy, CTAs are strategically placed within various types of content and communication channels to achieve specific goals. These goals can be part of a larger marketing, sales, or communication strategy. Here are a few examples of how CTAs fit into more complex strategies:

> **Content Marketing Strategy:** Within a content marketing strategy, CTAs are strategically placed within blog posts, articles, videos, or other content to guide the audience toward a desired action, such as downloading an eBook, subscribing to a YouTube channel, or visiting a product page.

> **Email Marketing Strategy:** In email marketing, CTAs are used to encourage recipients to take specific actions, such as clicking on a link to a product page, filling out a survey, or making a purchase. The overall email marketing strategy may

involve segmenting the audience, crafting personalized messages, and optimizing the timing of email sends.

➢ **Sales Funnel Strategy:** In a sales funnel, CTAs are used at different stages to move potential customers through the buying process. This could include CTAs to capture leads, nurture prospects, and convert them into paying customers. The sales funnel strategy may involve lead scoring, drip email campaigns, and retargeting ads.

➢ **Website Conversion Strategy:** On a website, CTAs play a crucial role in converting visitors into customers or leads. The website's overall strategy may involve optimizing user experience, A/B testing different CTAs, and tracking user behavior to improve conversion rates.

➢ **Social Media Strategy:** CTAs are often used in social media posts to encourage engagement, such as liking, sharing, commenting, or clicking through to a website. The social media strategy may include content planning, audience targeting, and measuring social media ROI.

In all these cases, the CTA is a tactical element that supports the broader strategic goals of the organization or campaign.

Effective CTAs are carefully crafted to align with the strategy and drive desired outcomes, and they are often tested and refined to improve their effectiveness.

The strategy that comes before a CTA (Call to Action) is typically the development of your marketing or communication strategy. Here's a simplified breakdown of the process:

> **Marketing/Communication Strategy:** Before you create a CTA, you need to have a clear strategy in place. This strategy should outline your overall goals, target audience, messaging, and the channels you plan to use. It should also consider your competitive landscape, budget, and timeline.

> **Audience Segmentation:** Part of your strategy involves identifying your target audience or audiences. You should know who you're trying to reach, what their needs are, and where they can be found.

> **Messaging and Content:** Based on your strategy and audience research, you'll develop the messaging and content that will resonate with your target audience. This includes creating valuable and relevant content that will lead up to your CTA.

- ➤ **CTA Development:** Once you have the right messaging and content in place, you can then create your Call to Action. This is the specific action you want your audience to take, such as signing up for a newsletter, making a purchase, or contacting your business.

- ➤ **Placement and Design:** Your CTA needs to be strategically placed within your content, whether it's on a webpage, in an email, on social media, or within a video. The design and visibility of the CTA also play a crucial role in its effectiveness.

- ➤ **Testing and Optimization:** After launching your CTA, it's essential to continuously test and optimize it. This involves A/B testing different versions of the CTA, analyzing its performance metrics, and making improvements based on the data.

- ➤ **Conversion Funnel:** Your CTA is a key part of your conversion funnel. It's where you guide your audience from awareness to action. Ensuring a smooth and logical progression through the funnel is vital for success.

- ➤ **Monitoring and Analysis:** Monitor the performance of your CTA and the overall strategy. Analyze data to understand how well your CTA is working and whether it's achieving your desired outcomes.

> **Iterate and Refine:** Based on the insights gained from monitoring and analysis, make iterative changes to your strategy, content, and CTA to continually improve results.

So, the strategy that comes before a CTA involves careful planning, audience research, content development, and positioning to ensure that your CTA is as effective as possible in achieving your marketing objectives.

www.ingramcontent.com/pod-product-compliance
Lightning Source LLC
Chambersburg PA
CBHW062329290526
45794CB00005B/1957